CHICKEN
Nuggets

CHICKEN
Nuggets
A miscellany of poultry pickings

DAVID SQUIRE

green books

CONTENTS

INTRODUCTION

Golden Sebright Bantams

Few other animals are such important food providers as chickens – daily and for people throughout the world. Unfortunately, their adaptable and amenable natures have been hijacked by many food manufacturers, making them victims of their own success. Commercially, chickens have often been constricted in conditions that are both psychologically and physically unacceptable in a civilized society. Encouragingly, however, many animal welfare organizations have, during recent decades, campaigned strongly to improve the welfare of commercially kept hens, as well as to rehome those that have come to the end of their commercial egg-laying life and can be given a more pleasant existence for a few years.

Shoppers have also played a vital role in improving the lives of commercially kept hens by only buying eggs clearly designated to have come from hens kept in humane ways.

This book is both a tribute to chickens and an insight into their lives, their pecking order, vision, language and even their love life. It also reveals ways to keep them, even in small gardens. They make superb family pets, with some even answering to their names.

More than most animals, chickens are enshrined in folklore, superstitions, proverbs and general sayings, which, perhaps unknowingly, can be heard almost every day. This small but information-packed book reveals many of them that will make you chuckle.

CHAPTER ONE

Where did Chicken/ Originate?

IN THE BEGINNING

About 10,000 years ago, stirrings among jungle fowls in South-East Asia triggered the development of the wide range of chickens we know today. It could never have been envisaged then how this would influence our eating habits, with today's global estimate of 27 billion chickens in existence at any given time.

Grey Jungle Fowl

EARLY BIRDS

The ancestry of modern-day domesticated chickens was earlier thought to have been both the Red Jungle Fowl (*Gallus gallus*) and Grey Jungle Fowl (*Gallus sonneratii*). Genetic research, however, now indicates that the Grey Jungle Fowl is the most likely ancestor, although some authorities suggest a more complex origin.

Multiple origins in distinct and separate areas of South and South-East Asia, as well as India and

North and South China, have also been suggested.

FOOTLOOSE FELLOWS
From the area now known as Vietnam, chickens spread to India, with their domestication reaching Asia Minor and Greece some 7,000 years ago. They also spread from Thailand to China by 5,000 BCE, and thereafter to Japan.

By the 18th Dynasty (1550–1292 BCE) chickens had been introduced into Egypt, resulting in a doorway to Europe. Nowadays, chickens are kept for their eggs and meat in every part of the world except Antarctica.

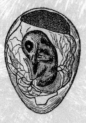

CHICKEN OR EGG?
The schoolboy riddle *'Which came first, the chicken or the egg?'* was resolved in 2010 by two British universities using a super-computer. The research concluded *'The chicken came first'*.

CHICKEN MYTHOLOGY
In many cultures, chickens are sacred animals and part of religious worship:

❧ Cock-birds were known as 'Persian Birds' to early Greeks because of their importance in Persian religious worship.

❧ In Indonesia, chickens were of importance during Hindu cremations, being an escape channel for evil spirits present in the ceremony.

❧ A fight between two cock-birds was essential at Balinese festivals or religious ceremonies.

❧ In the 6th century, Pope Gregory 1 (Gregory the Great) declared a cock-bird to be an emblem of Christianity.

❧ The Roman general Publius Claudius Pulcher had his chickens thrown overboard when they declined to feed before the Battle of Drepana (249 BCE), claiming *'If they won't eat, perhaps they will drink.'* He lost the battle and 93 of his 123 ships were sunk!

CHICKENS IN BRITAIN

Chickens were established in Britain before the arrival of Julius Caesar in 55 BCE and are thought to have been spread much earlier by Phoenician traders who sailed along the Mediterranean, around the Iberian peninsula, then north to Ireland and Britain. Chickens also arrived from Europe.

BREEDS

❧ **Cornish** Also known as Indian Game, with two varieties – Cornish Game and Jubilee Cornish Game (lighter and less stocky). Originally, the breed was used for cock-fighting, but it is an excellent meat bird. It is seen in several distinctive colours – Dark, White, White-laced Red, Buff, Blue-laced Red, Spangled, Black and Mottled.

❧ **Old English Pheasant Fowl** Although it has 'pheasant' in its name, it is not a pheasant, gaining this name solely because of its resemblance to that type of fowl. Earlier, it was known as the Copper Moss, Golden Pheasant, but since 1914 has been called the Old English Pheasant Fowl.

Buff Orpington

❧ **Orpington** There are several popular varieties, all bred by William Cook and named after the then village of Orpington in Kent, England. In 1886 he introduced a black variety, a white type in 1889, and a buff form in 1894.

❧ **Red Dorking** The original Dorking is thought to have been introduced from Italy to Britain by the Romans and was said to be the first epicurean chicken in England, with capons raised for banquets. It was mentioned during the time of

Julius Caesar (100–44 BCE) by the Roman agricultural writer Lucius Junius Columella.

Red Sussex

❧ **Sussex** This is claimed to have been in existence for more than 2,000 years, mainly kept for its meat, and to have derived from the Ardennes region of Belgium. The oldest variety of the Sussex breed is the Speckled; later, Brahma, Cochin and Silver-grey Dorkings were used to produce the now-famous Light Sussex.

❧ **Scots Dumpy** A breed with a long heritage; it is said to have been known in the 11th century. It has a squat-like appearance and tends to waddle with its body low to the ground. Dumpies were earlier known as Crawlers, Creepies, Bakies, Dadies, Hoodies and Stumpies. It is claimed that the Picts (a race of people who inhabited northern Britain in the

9th century, later uniting with the Scots) took advantage of this breed's acute hearing by using it to warn of invaders treading on thistles. Legend suggests that this is why the thistle was adopted as a Scottish emblem.

❧ **Scots Grey** Earlier known as the Scotch Grey, it was recorded as a barnyard breed in the 16th century. It is thought to feature in the ancestry of the Dorking and has a coat similar to the Barred Plymouth Rock and Cuckoo Maran.

Scots Grey

BEGINNING OF POULTRY CLUBS

The Poultry Club of Great Britain was founded in 1877 to safeguard the interest of all pure-bred and traditional breeds of poultry, including chickens, ducks, geese and turkeys.

CHICKENS IN EUROPE

As chickens spread from Asia to Europe they were crossed and re-crossed several times to create a wide spectrum of breeds, which themselves were used to produce further types. Many were introduced into Britain and North America, then to other countries, including Australia and New Zealand.

Ancona

BREEDS

❧ **Ancona** Originating in Italy and earlier called the Marchegiana, it was taken to England in the mid-1800s from the sea port of Ancona, from where it derives its name. It arrived in North America in 1890. Its black plumage is made even more distinctive by white 'V' tips at the ends of its feathers.

❧ **Andalusian** An old Mediterranean breed originating in Andalusia, a large area in the south of Spain. Ofte known as the Blue Andalusian, it does not always breed true to that colour and has been used in genetic experiments (see page 91).

Appenzeller Spitzhauben
Originated during the 16th
century in the Appenzell region in
north-east Switzerland. The word
'Spitzhauben' means pointed hood
and this is recognized in the breed's
crest, a style claimed to derive from
the fancy and fluffy hats worn by
women in that region. The breed
is acknowledged as the national
chicken emblem of Switzerland.

Barnevelder A popular breed,
with its origins in the Barneveld
area of the Netherlands, it made
its debut in 1911 from a medley
of Asian breeds introduced into
that area between 1850 and 1875.
Brahma, Cochin and Malay fowls
were crossed with local chickens.

Crèvecœur One of the oldest
French breeds and shown in Paris
in 1855. First known as the Black
Polish, it originated in the village
of Crèvecœur in Normandy from
a mixture of the Polish breed and
other fowls. It is claimed to be in
the ancestry of La Flèche, Houdan
and Faverolles breeds.

Houdan Earlier known as the
French Dorking because of its fifth
toe, it was recorded in the 17th
century but mainly developed in
the mid-1800s. It is well known for
its large headdress and beard.

It arrived in England in 1850 and
was known to be in North America
in 1859. It is named after the city of
Houdan, near Paris.

La Flèche Originating in
the Le Mans area of France, with
Black Spanish, Crèvecœur, Black
Polish and Minorca genes in its
development. As a result of the
excellent flavour of its white breast
meat it became popular in Paris
markets in the mid-1800s; its
popularity declined in the mid-1900s
but is now making a comeback.

Leghorn The breed originated in
Italy, with birds being exported from
the port of Leghorn on the west
coast. It was introduced into Britain
in the mid-1800s and to North
America in 1835, where the breed
was initially known as 'Italians'.

Polish A distinctive breed
sometimes known as the Poland, it
is characterized by its turban-like,
rounded and crested head feathers.
Known in Holland in the 16th
century, the breed is claimed to
have originated in an area around
the Baltic states, a region then
controlled by Poland. It arrived in
England in 1816 and was shown at
a poultry show in 1845.

CHICKENS IN NORTH AMERICA

North America has been the cradle for many superb breeds that are now world-famous. Chickens were taken from Europe to North America by the Pilgrim Fathers and other early settlers, but later developments were mainly the result of using the genes of more recently imported fowls, often Asian in heritage.

BREEDS

❦ **Buckeye** Developed in 1896 in Warren, Ohio, and claimed to be the only breed created by a woman, Nettie Metcalf. It is a handsome chicken, refined from several breeds including a Buff Cochin, Barred Plymouth Rock and Black-breasted Red Game. Because the breed has a pea comb it was originally known as the Pea Combed Rhode Island Red, but changed in 1902 to Buckeye. The name Buckeye is derived from Ohio's nickname, 'Buckeye State' (a reference to the Ohio Buckeye, a species of Horse Chestnut).

❦ **Dominique** Earlier known as Pilgrim Fowl or Dominicker, it is considered to be the oldest American breed of chicken and to have been introduced into America by the Puritans. It is a handsome breed, with its genes in the parentage of the Plymouth Rock. In the 1900s its numbers declined, but it is now rightfully receiving a revival of interest.

❦ **New Hampshire Red** A placid and friendly breed with the Rhode Island Red in its ancestry. Through selective breeding it originated in New Hampshire, the Granite State, in the early 1900s and was recognized as a distinct breed in 1935.

Plymouth Rock

❦ **Plymouth Rock** Often thought to be the most popular breed in North America; it was first shown in 1869 and claimed to be a cross between several breeds, including the Cochin, Brahma, Java, Dominique

Golden Wyandotte

and Minorca. The breed is credited to the Reverend D. A. Upham of Worcester, Massachusetts.

❧ **Rhode Island Red** Beautiful plumage, and originally developed to withstand cold winters yet at the same time to lay plenty of eggs. Its development was in the town of Little Compton, Rhode Island, in the 1800s, and it was first exhibited in Boston in 1880. Genetically it is a mixture of the black-red fowls of Shanghai, the Malay and Java. It is the official state bird of Rhode Island.

❧ **Wyandotte** The breed originated in the 1870s in North America in the form of the Silver-laced Wyandotte; it was recognized as a new breed in 1883. Other varieties followed and now there are 17 distinct colour forms.

PRAIRIE CHICKEN

The Greater Prairie Chicken is a large bird in the grouse family and native to North America. Once abundant, it is now rare or extinct over much of its earlier range in tall-grass prairies. Severe winter weather does not usually reduce its numbers, but it is Man who has mainly decimated it by turning the native prairie into cropland. There are now active conservation measures to aid the breed's re-establishment.

Greater Prairie Chicken

CHICKENS IN SOUTH AMERICA

Exactly how chickens were introduced into South America is under debate, with thoughts of a Columbian origination as well as the Spanish Conquistadors being the main avenues. A case is also made for chickens having been taken, perhaps by Polynesians (or the Chinese), across the Pacific from Asia.

THE CHINESE QUESTION

In his book *1421: The Year China Discovered the World*, Gavin Menzies suggests that within the years 1421–3 Chinese ships visited both the east and west coasts of North and South America, taking with them plants and animals, including chickens.

When Ferdinand Magellan and the Spanish Conquistadors reached South America, the chickens they found did not resemble European breeds. Instead, they encountered, among others, a breed similar to the Asian Frizzle, and tall and thin Malay fowls. These are claimed to have been the result of Chinese introductions many years earlier.

Black Araucana

SOUTH AMERICAN BREED

❧ **Araucana** Known in North America as the South American Rumpless, it was raised in the Araucana region of central Chile by the Araucana Indians. It lays blue eggs (a characteristic of many Asian breeds) and is said to have descended from wild jungle fowl known as the 'chachalaca', a breed thought still to exist in the Amazon Basin and isolated places in the Andes.

❧

NORTH AMERICAN VARIATION

❧ **Ameraucana** A breed developed in North America from the Araucana. It is a mixed and non-standard breed with imprecise and variable traits, but it does provide colourful eggs – shades of light blue and green!

It is a larger and heavier breed than the Araucana, with better meat quality. Additionally, it differs from the Araucana in that it reveals a tail and beard, but no ear-muffs.

CHICKENS IN THE ANTIPODES

In earlier centuries, chickens were one of the few food animals that could be confidently taken on long sea trips, providing food on the journey as well as breeding stock on arrival. Early breeds of chicken taken to Australia were traditional English types, including the Sussex.

LONG PENAL PADDLE

There are records of chickens being sent from England to Australia in 1787 in a fleet of 11 ships carrying 759 convicts. They were accompanied by sheep, pigs, goats and horses. Establishing a new colony and home for people was difficult and uncertain, but it is certain that the scavenging nature of chickens helped in their own survival as well as that of the colony.

~

POPULAR AUSTRALIAN BREEDS

Australia has produced several superb breeds to suit its climate.

❋ **Australorp** Sometimes known as the Black Australorp, it was developed in Australia from Black Orpingtons. Opinions differ about the name Australorp, with claims that it derives from the Australian Black Orpington.

It is a breed famed for its egg-laying qualities. In 1922–23, six hens set a record for laying eggs, with an average of 309 eggs per hen over a period of 365 days.

❋ **Australian Langshan** A smaller version of the Croad Langshan. Croads originated in the Langshan District in China and were taken to England in 1872 by Major F. T. Croad. They are well suited to the warm, sunny climate of Australia.

❋ **Australian Game Fowl** Earlier known as the Colonial Game and developed to provide fighting cocks and entertainment for soldiers guarding convicts. Usually known as 'Aussies', these birds have an Asiatic heritage, with a tall, leggy outline and a long neck.

Plymouth Rock

NEW ZEALAND SPECIALITIES

Many breeds are popular in New Zealand, including the Light Sussex, Leghorn, Plymouth Rock, New Hampshire Red, Orpington and Wyandotte. But there are also excellent hybrids specially developed for their egg-laying or meat qualities.

CAPTAIN COOK LEGACY

On his second voyage to New Zealand in 1773, Captain James Cook gave hens to the Maoris in both the North and South Islands. Later, in 1814, missionaries in the Bay of Islands, towards the top of the North Island, became the first recorded poultry farmers in New Zealand.

Early settlers regarded eating chickens as a luxury, to be indulged in perhaps only once or twice a year, and even then only a hen that had come to the end of its egg-laying life. Nowadays, chickens (often known as 'chooks') are kept by almost half of all households in New Zealand.

CHICKENS IN ASIA

Asia is the birthplace of domestic chickens. As chickens spread westward into Europe, later to North America and the Antipodes, breeds were developed primarily for their meat and egg-laying abilities. Asia, however, retained some of the most decorative and flamboyant of all breeds.

Onagadori Because it takes more than three years for the tail feathers of this distinctive breed to moult, tail lengths of up to 9 m (30 ft) are possible, although they are usually shorter. The breed originated in the Kochi Prefecture of Japan and is so respected in Japan that it has been granted preservation status by the government.

Sultan

There are several colour variations, including the Black Breasted Red, Black Breasted Silver, Black Breasted Golden, and White.

Sultan Originating in Turkey and particularly bred for the entertainment of sultans, they were known as 'Fowls of the Sultan' and 'Moving Flowers'. Occasionally, they were called the 'Polish Fowls of Turkey'. So treasured were they that only sultans and the ruling class were allowed to keep them. It was not until 1854 that they arrived in Britain, and 1867 in North America.

They are dainty and likened to powder-puffs on legs, with profuse head decoration, puffy crests, beards, long tails and foot feathering.

Sumatra Native to the island of Sumatra and parts of Malaysia, this breed was earlier known as the Sumatra Game Bird, Sumatra Pheasant, Black Pheasant and Java Pheasant Game Bird; it is considered to have been bred directly from jungle fowl and wild pheasants. Males have long, cascading tails.

In the mid-1800s it was introduced into Europe and North America, primarily for cock-fighting, although now usually kept for exhibition.

Yokohama Like the Onagadori, this is a long-tailed breed. It did not originate in the Japanese city of Yokohama as its name suggests, but it was from there that French missionaries first exported it to Europe. Tails are long, often 90 cm (3 ft), much shorter than in the Onagadori, but nevertheless highly attractive.

The breed's origination is complex and owes its current appearance to two Japanese breeds – the Onagadori and Minohiki (Saddle Dragger).

Black Sumatra

CHAPTER TWO

Getting to Know Chickens

IDENTIFYING CHICKENS

Chickens receive more publicity each year than any other food animal and therefore it is not surprising they are easy to recognize. At first glance all chickens may appear the same, but there are wide variations in sizes, shapes and colours. Also, hens and cock-birds differ markedly.

PHYSICAL APPEARANCE

The range of chickens is extensive, a result of dedicated breeding in many countries to produce a chicken to suit local conditions and culinary demands. But they all have the same basic structure.

❧ **Heads** are usually small and have strong beaks.

❧ **Bodies** are covered in feathers (although Israeli scientists have bred a featherless chicken, which is claimed to grow faster than a feathered type and to produce low-fat meat).

❧ **Wings**, well covered in feathers, are usually short and not adapted to flying, although a few light breeds have limited flying abilities.

❧ **Breast bones** are shaped liked the keel of a boat, resulting in great strength.

❧ **Tails** feature on most chickens, with the exception of the Araucana

Hen bird

Cock bird (rooster)

and a few other 'rumpless' breeds. Most tails are relatively small and erect, while a few, such as those of the Japanese Onagadori, are long and flamboyant.

❧ **Combs** are apparent in most chickens, with some being distinctively flamboyant (a range is shown on pages 46 and 47).

❧ **Strong, upright and scaly legs** enable rapid mobility. With most breeds, the legs are free from feathering, but a few, such as the Cochin, are completely covered in feathers, forming decorative frills.

❧ **Strong and evenly spread toes** are vital for balance, running and perching; most breeds have four toes, some five.

❧ **Claws** are strong and for defensive purposes, although cock-birds use them to cling to and control a hen when mating.

⌇

HENS & COCK-BIRDS

The usual and obvious difference is that male birds are larger, more dominant and aggressive than females. Additionally, cock-birds have more flamboyant tail feathers and a larger and more colourful and distinctive comb.

Cock-birds are usually noisier than hens and this may be a problem if you live in suburbia. Fortunately, however, male birds are not essential in your group of hens for the production of eggs.

CHICKEN TERMINOLOGY

All groups and types of animals have a vocabulary to describe their appearances and activities – and chickens are no exception. In addition to descriptive terms, chickens have many words related to their culinary qualities and uses. Here are a few terms that seek to define chickens.

GENDER TERMINOLOGY

* **Young males and females** Widely known as chicks.

Hen

* **Females** Young ones, up to the age of about a year and before starting to lay, are pullets. After that they reach adulthood and become hens.

* **Males** Up to the age of about a year, a male bird is known as a cockerel. After that stage he becomes a cock, cock-bird or rooster, depending on the country.

* **Castrated males** Known as capons. Traditionally, castration was undertaken surgically, later chemically, but both are now banned. Surgical castration is cruel to the bird, while chemical castration puts human eaters at risk from absorbing female hormones.

~

BOILER OR BROILER?

These sound similar, but although both relate to culinary matters, they are quite different:

* **Boiler** An old hen, perhaps earlier kept for its ability to lay eggs, which has now radically diminished. When killed for eating, it requires long and thorough cooking. Nevertheless, it makes an excellent meal.

* **Broiler** A young bird (usually less than eight weeks old) specially raised to be killed for its meat.

Silver Spangled Polish

A QUESTION OF BREEDS

There are many breeds of chickens to choose from, as well as sizes:

* **Pure-breds** These are breeds that are well established and, when members of the same breed are bred together, the purity of the breed is maintained. They usually lay eggs for more seasons than hybrid types, but are more likely to become broody and stop laying for a time.

* **Hybrids** These are the progeny of crosses between two pure-bred breeds. The result is a showing of hybrid vigour in their offspring. However, hybrids do not reliably pass on their features and characteristics to their progeny.

They were mainly developed in the early 1950s in response to an increased demand for eggs, but their egg-laying abilities often decrease after their first season.

* **Large-fowl breeds** Sometimes known as 'standard-sized' breeds, they represent the normal size to be expected for a particular breed. Within this group there are 'light' and 'heavy' breeds.

* **Bantams** These are smaller than large-fowl breeds, and are usually inquisitive and amusing characters. They are often kept as family pets. Within this group there are 'true bantams' (small and unrelated to large-fowl breeds) and 'miniatures' (small versions of large-fowl breeds).

White Rosecomb Bantam

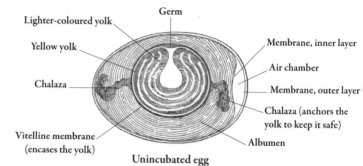

Germ

Lighter-coloured yolk

Yellow yolk

Chalaza

Vitelline membrane
(encases the yolk)

Membrane, inner layer

Air chamber

Membrane, outer layer

Chalaza (anchors the
yolk to keep it safe)

Albumen

Unincubated egg

BROODY HENS

Broodiness is a natural and inherited function in hens to ensure the development of another generation of chickens.

Ancestors of modern chickens laid two clutches of eggs each year – one in spring and the other in late summer. This reproductive characteristic has almost been eliminated from modern breeds, especially those raised for commercial egg production. Nevertheless, it is still present in some heavy breeds.

ALL ABOUT EGGS

A chicken's egg is a miracle of nature and one that has been hijacked from its reproductive role to feed millions of people each day. Additionally, medical research indicates that eating eggs may prevent age-related macular degeneration in eyes. Another bonus is their use in weight-loss diets.

❧ Depending on its size, each egg provides 60–80 calories.

❧ An average-sized egg weighs about 57 g (2 oz); the shell forms 11% of this, the yolk 31% and the white 58%.

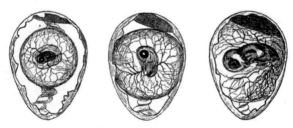

Embryo at 5, 8 and 11 days

RESCUED HENS

An awareness of the dire plight of hens kept in battery conditions has become better known in recent decades, with specialized animal welfare associations rescuing hens at the ends of their egg-laying years and rehoming them.

These ex-battery chickens (known as 'spent' or 'ex-bat' hens) would have lived in commercial egg-producing establishments until about 72 weeks old. Their fate, if they had not been rescued, would have been to be killed, with their meat used in lower-grade meat products and as pet food.

For some 'spent' hens the future is happier: to be rehomed in groups of three or more as family pets. In the British Isles well over 300,000 have been rehomed in the last five years, with the British Hen Welfare Trust, the largest hen welfare organization, especially active in this humane endeavour. Additionally, Compassion in World Farming, founded in 1967, is at the forefront in fighting for the better treatment of farm animals.

CHICKEN PSYCHOLOGY

Chickens have been popularized as 'bird-brained' creatures with limited intelligence. However, this is far from the truth as recent research indicates that they are bright animals and able to resolve complex problems. Additionally, they reveal self-control and worry about their future.

SO WHAT CAN THEY DO?

In many ways, chickens are more intelligent than cats and dogs – and even some primates when young. They are able to:

❧ Pass information from a mother hen to her chicks.

❧ Understand cause-and-effect and be aware that objects still exist even when hidden from view. In this respect they are cleverer than many small children, who usually disregard hidden objects.

❧ Have a social structure known as the pecking order (see pages 30–33). This helps to prevent squabbles and irritations occurring between chickens – both cock-birds and hens.

❧ Recognize more than 100 other birds and relate them to their own social hierarchy.

of life. And should a mother hen not be present (when eggs are being hatched in an incubator), a chick will imprint with you and consider you to be its 'leader'.

✤ Where young chicks have been bought from a supplier, they will never imprint with you in the same way as home-hatched chicks. But they will recognize you in their pecking order.

CHICK FEAR

Within the third day of hatching a chick starts to become fearful. Normally, the mother hen will calm it, but if you have become the surrogate mother try talking softly to it – a gentle lullaby is superb! Essentially, avoid making loud and sudden noises and movements.

✤ Have more than 24 individual vocalizations, which mother hens start to teach their chicks at an early age. Some of these are even taught to chicks before they hatch.

✤ Be highly protective of their 'ladies' – cock-birds will endeavour to keep them safe even during times of dire threat.

✤ Be quicker than dogs (even the super-intelligent border collie) at responding to sound information, especially when reacting to 'clicker' training techniques.

CHICK IMPRINTING

Most animals are able to identify and imprint with their own family – and chicks are no exception.

✤ Chicks start imprinting with a mother hen during their first day

THE PECKING ORDER ~ & YOU!

To have harmony within a group of animals there needs to an accepted hierarchy. In 1921, the Norwegian zoologist Thorleif Schjelderup-Ebbe (1894–1982) wrote a PhD dissertation about the pecking order of chickens, partly based on his observation of them since the age of ten.

THE 'PECKING ORDER' CONCEPT

Also known as the 'peck order', this initially referred to chickens, but was later adopted in the world of finance and general business.

With chickens, it establishes which is the 'top chicken' and the 'bottom chicken', and where the rest fit in between. The dominance level determines which chicken gets preferential access to food and mates. It also helps to reduce discord within a group.

The 'pecking order' within groups of animals can be interpreted as being brutal, but ensures that the healthiest and most dominant male controls the pack. This enables the best characteristics within a group of chickens to be passed on to the following generation.

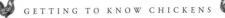

Charles Darwin (1809–1882), the world-famous English naturalist, would have agreed that the 'pecking order' concept substantiated his scientific proposal of natural selection, as explained in his book *On the Origin of Species* (1859).

Charles Darwin

THE 'PECKING ORDER' — & YOU!

The routine of feeding and caring for your group of chickens soon establishes you as the leader of their flock. Depending on their gender, they will approach you in different ways. For example:

Cock-birds Initially, a cock-bird will approach you cautiously, waiting to see what you are about to do. He is the leader of his group, protector of his hens and therefore cannot trust anyone until he knows them.

He will slowly and cautiously approach you, staring directly at your face and awaiting your reaction. Then he will slowly flap his wings, opening them to give the appearance of greater size.

This hesitant and initially aggressive attitude will fade as soon as he knows you are a friend and will present no threat to his hens.

Hens These may initially greet you in the same way as a cock-bird, but soon become friendly when they realize you have brought food for them. They are usually eager to see you and, when your relationship with them is established, will eagerly await your arrival.

Some hens will stand on your shoes and peck at your trousers; this may indicate that they wish to be picked up and stroked. Take care, however, as this may lead to jealousy among the other hens. Always treat all of them in the same way.

PECKING-ORDER DEVELOPMENT

This begins quite early in a chicken's life.

❧ **From the age of two weeks**, chicks tend to peck at each other, initially playfully, but soon some of them become excessively confident in their abilities; other chicks will become wary of them and fall down the pecking order.

❧ **When nearing puberty** (at 10 to 12 weeks of age for pullets and 12 to 16 for cockerels), the pecking intensifies and becomes especially aggressive when settling disputes over food and territory. Although it may appear brutal, never interfere with this process.

❧ **At about 26 weeks**, the birds usually settle down and accept the eventual pecking order.

❧ **Where there is only one cock-bird** within a group of chickens, it becomes dominant and is accepted as the top animal.

❧ **When a cock-bird is established** in his role of top animal, hens start to compete with each other for his attention. This can become aggressive and exceptionally vicious, with each hen trying to diminish the role of other hens.

❧ **If a group is formed solely of hens** they battle to find the dominant one. Often the hen with the largest comb succeeds in getting the top job. And when established, she may even start to crow; also, she may try to sexually mount other hens.

❧ **If a further cock-bird is introduced** into a group with an established male, mayhem inevitably results. Discord usually spreads to the entire group, with fighting between the two males.

❧ **Disagreements between hens** can become as ferocious as when they involve two cock-birds. Do not use bare hands to part them; rather, gently spray them with clean water. This will divert their attention and they can be separated, using gauntlets.

❧ **When a cockerel is introduced** into a group of hens, he may not be accepted and the top hen will fight him. Additionally, it is doubtful if he would be allowed to sexually mount a hen; inevitably, he becomes a sad and dejected fellow that, eventually, has to be put in a different pen.

❧ **When introducing new hens** into a long-established group the reaction can be unpredictable. Here are a couple of solutions from folklore. Wait until dark and quietly put the newcomer into the pen. Another way to help is to rub garlic over the hen that is to be introduced. Cut a clove in half and rub it over the bird's chest.

VISION

Few creatures on this planet have such good vision as chickens and other birds. A chicken has an eye on both sides of its face, enabling it to see in three dimensions as well as having all-round vision. This is a major advantage for detecting predators, especially in the wild.

COLOUR VISION PAR EXCELLENCE

The development of colour vision in animals has varied from one species to another, with some having exceptionally sensitive and responsive sight. This reflects how all species adapt to their environments and thereby survive.

🌿 **The human retina** is able to detect red, blue and green wavelengths, whereas chickens extend this to violet wavelengths, including some ultraviolet. Additionally, they have receptors called double-cones that aid in the detection of motion, alerting them to danger.

The colour receptors in the eyes of chickens are far more numerous than in mammals, giving them superior vision. A chicken is ideally equipped to identify colourful plumage as a way to find a suitable mate. They can also quickly identify colourful berries and fruits from a distance.

DINOSAUR LEGACY

Part of the ability of chickens and other birds to see exceptionally well is thought to be the result of spending a long part of their evolution in the light.

During the age of dinosaurs (to whom birds are related) most mammals became nocturnal to ensure their survival. Birds did not suffer from this evolutionary problem of having to live in the dark and therefore now have better vision.

FEEDING & FORAGING TECHNIQUES

When searching for food, a chicken first focuses with one eye, then the other. This explains why head movement is rapid, from side to side, as a chicken walks looking for food.

A chicken normally detects a grain of food from about 90 cm (3 ft). When about 10 cm (4") away it tilts its head, leans downwards and eats it.

CHICKEN-SPEAK

All animals communicate with each other. Domestic chickens are said to have a greater range of communication sounds than birds in the wild – perhaps the result of their long domestication, about 10,000 years – and are able to produce more than 24 different sounds.

A QUESTION OF HEARING

Chickens have a well-developed sense of hearing, although initially this may appear to be a false claim as they do not have ears. This is because their auditory canals are hidden behind a flap of skin and feathers. Nevertheless, their sound receptors resemble those in humans and are formed of three basic parts:

❧ **The outer ear** is a tube leading to the ear drum (tympanum).

❧ **The middle ear** has a single bone stretched across it, which is known as the columella.

❧ **The inner ear** is bathed in fluid (the outer and middle parts are filled with air).

❧

INTERPRETING SOUNDS

Chickens, as well as other birds, hear sounds differently from humans. Whereas birds recognize sound in something similar to 'perfect' pitch (sometimes known as 'absolute' pitch), humans detect sound in 'relative' pitch. This means that if humans receive sounds in one octave, they are able to recognize it in a different octave.

Chickens cannot do this, although they can recognize a fundamental note combined with harmonies, which enables them to respond and imitate it – in several ways.

❧

BEING LIKE DAD

Male birds are famed for their 'cock-a-doodle-do' and are able to do this from the age of three to four months. Initially they begin with a few rather subdued squeaks.

❧ **The volume of their calls** is influenced by the bird's size and body weight.

❧ **Some cock-birds** are able sustain a 'cock-a-doodle-do' for several minutes, but usually it is just a short series of notes.

❧ **If angered or threatened**, a cock-bird produces a low-pitched note to warn hens of approaching danger.

❧ **A cock-bird** captures the attention of hens by emitting a high-pitched but friendly and non-aggressive note.

❧

LEARNING THE LANGUAGE

Chicks, hens and cock-birds have their own languages, which are influenced by the onset of hatching, proximity to danger, the presence of food and relationships with members of the opposite sex.

Chick talk Even before hatching, a hen makes contact with her chicks and is able to offer them comfort and reassurance. Then, as a chick 'pips' on the inside of an egg and emerges, communications increase.

Chicks cheep, while mother hens cluck and, if separated, a chick begins a long, high-pitched note of panic. The higher the pitch from the chick, the more rapidly a mother responds, usually with a deep, soft voice that provides reassurance. This can occur over a distance of 9 m (30 ft) or more.

Contented chicks, when fed and allowed to slip under their mother's feathers, reveal a soft chirp. They communicate with each other, but do not go to their assistance should they be at risk or get lost.

Contented hens When happy and not in danger, a hen reveals an even and rhythmic cluck, but if alarmed, this turns into a jumbled 'gagaga' sound.

When a hen is about to lay an egg she produces a laboured, almost complaining tone. Afterwards she resorts to a cackle of contentment.

❀ **Feeding information** A chicken will make a 'tck, tck, tck' note when in the presence of food. However, a cock-bird will also make this noise upon discovering anything edible. This encourages hens to join him (which can be a prologue to sex).

❀ **Asiatic chickens** are claimed to make different sounds from those of European breeds. Instead of 'cock-a-doodle-do', it more closely resembles 'kik-kiri-kee'.

39

CHICKENS IN LOVE

Chickens mainly mate in the afternoon, when there is the best chance of producing a fertilized egg. Nevertheless, they are sexually active throughout daylight hours, with many poultry experts reporting that cock-birds eagerly mate with any hen that cannot run away fast enough!

COURTING & SEXUAL RITUAL

A cock-bird performs a distinctive ritual when wooing a hen:

✤ **He entices a hen** by leading her out to food, clucking in a high pitch and allowing her to feed first.

✤ **Hens reluctant to indulge** in sexual activities escape by running away and moving their tails from side to side; they may also just crouch.

✤ **If a hen stands still**, the cock-bird stands in front of her and fluffs out his neck feathers. He then dances around her, dragging the wing closest to her on the ground. This part of his ritual is often known as the 'waltz' or 'wing-fluttering' stage.

✤ **He then pounces**, jumping on her back and pecking at her neck feathers or skin to retain balance. He uses his feet to push her down. This is often known as 'treading'.

✤ **The hen lowers her head** and raises her tail (an indication of receptiveness) in readiness for the sexual act.

✤ **Hens and cock-birds** are not endowed with sexual organs comparable to human ones and therefore it is necessary for the cock-bird to move his cloaca near the hen's cloaca and deposit his sperm inside her. This is called the 'cloacal kiss' and lasts about ten seconds.

✤ **The cock-bird ejaculates** between 100 million and 5 billion sperm at one time – the higher figure at the beginning of the day.

& AFTERWARDS...

❧ **The cock-bird jumps off** the hen, sometimes encircling her and crowing boastfully.

❧ **The hen initially stands up** and remains still, then shakes herself and moves away.

❧ **A cock-bird may mate** 10–30 times a day, depending on the number of hens available.

To restrain his sexual activities and to keep him healthy, he needs to be put in a separate pen for one or two days each week. Also, he should not have more than eight hens to look after (preferably six).

CHICKEN SEXUAL TERMS

The sexual act of chickens has gathered many terms, some archaic and now rarely used:

❧ **Flogging the hen**

❧ **Jumping**

❧ **Mounting**

❧ **Treading**

CHICKEN FEATHERS

Feathers grow from the outer layer of skin and form a bird's plumage; they create protection from rain, hot and cold weather and injury. Some colour styles within a feather appear only in male or female members of a specific breed, but they all help to readily identify chickens.

FEATHER TYPES

Feathers are complex and variable in structure and are known to have covered some dinosaurs.

❦ **Each feather has a central**, hard quill (known as a shaft) with barbs attached to it that form a web-like, vaned and smooth surface.

❦ **Feathers undergo wear** throughout their lives and are replaced periodically through moulting; new ones develop through the same follicles as the fledged ones. New feathers are known as blood or pin feathers.

❦ **There are two basic types of feather** – vaned feathers (which cover the main and exterior part of a chicken's body) and down feathers (which appear underneath the vaned type).

In addition to their weather-proofing properties, in many birds feathers are used to insulate nests to provide comfortable conditions for eggs and chicks.

FEATHERS & THEIR MARKINGS

There are many patterns and colours revealed in feathers, including:

Barred

Barred Horizontal stripes across the feather in two, irregular or regular, different colours.

Cuckoo Irregular light and dark bars, with a dark tip to each feather.

Laced Dainty, with each feather having a narrow band of colour around its outside.

Mottled Feathers with white tips at their ends; not every feather on a bird will reveal this characteristic.

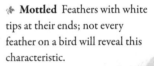

Pencilled

Pencilled Distinctive bars, some thin, forming concentric rows that closely follow a feather's shape. They are mostly seen in female birds.

Spangled Tips of feathers with contrasting white or black (or both) markings; these can be pear-shaped, resemble a half-moon or have a well-defined 'V'.

Splashed Irregular splashes of contrasting colours.

Stippled Feathers dotted in colours that create an attractive contrast to their background.

Striped Uniform colour along a feather's centre, surrounded by lacing.

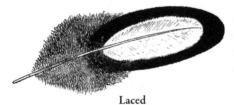

Laced

COLOURFUL PLUMAGE

The spectrum of colours and their patterns seen in hens and cock-birds is impressive, enabling them both to identify members of their own breed and attract a mate. The colour range is wide, from white to black and including red, blue and buff, with the addition of shades of yellow.

RANGE OF COLOURS

The attractive plumage of chickens is created from both a wide range of colours and the patterns seen in feathers (page 43). As chickens have evolved – and especially since chicken enthusiasts began selectively breeding them – a richer kaleidoscope of colours has evolved. This pageant has generated its own vocabulary, with some descriptions being particular to chickens and others more general and widely used.

❧ **Bay** Warm and golden-brown and particularly attractive when sunshine glances off it; even in winter it has a seasonal richness.

❧ **Birchen** Medley of black and pure-white, an allusion to the Silver Birch tree, which has silvery-white bark with irregular dark fissures.

❧ **Black** Lustrous greenish-black, a handsome colour and especially

revealed when in strong sunlight, particularly during summer.

❧ **Black-breasted red** Black everywhere, except for red hackles (on the neck), shoulders and parts of the wings. These colours partly mirror those in the Red Jungle Fowl.

❧ **Blue** Slate-blue, but variable and can include lavender. It is a restful colour and very attractive.

❧ **Brown-red** Birchen and a fusion of brown, black and orange.

❧ **Buff** Orange-yellow, with a rich golden undertone.

❧ **Chestnut** Rich, warm, dark reddish-buff.

❧ **Cinnamon** Dark reddish-buff.

❧ **Duckwing** Distinct bar across the wing of a male bird.

Fawn Light brownish-tan.

Mahogany Deep reddish-brown.

Partridge Patchwork of colour and resembling the patterns seen in partridges.

Porcelain Soft and gentle fusion, formed of straw-coloured feathers tipped in white with a pale blue stripe through part of them.

Quail Pattern of colours seen in quails, where black neck, back and saddle feathers are laced in golden-bay.

Red Dramatic and slightly variable colour, with a spectrum from rich dark red to mahogany-red.

Salmon Reddish- or pinkish-buff, said to resemble the colour of cooked salmon.

Self-colour Having plumage in a single colour.

Silver-pencilled Pattern formed of silvery-white and pencilled feathers (see page 43 for details of pencilling).

Slate Dark and bluish, sometimes appearing nearly black.

Wheaten Colour of wheat when seen in strong sunlight and sometimes resembling yellowish-ochre.

White Feathers entirely white.

Willow Yellowish-green.

RANGE OF COMBS

Combs, which are colourful and often intricately shaped, adorn the heads of chickens; the only exceptions are exhibition breeds and game fowls 'dubbed' for show – a somewhat barbaric practice in which the combs of fighting cock-birds are trimmed to decrease the surface areas open to attack.

UNDERSTANDING COMBS

They have several purposes in the lives of chickens:

☙ **They enable chickens to identify each other** and enable males to attract females.

☙ **Shapes and sizes vary** from one breed to another, with male birds having larger and usually more flamboyant combs than those on females.

☙ **They are fleshy, soft and 'alive'**, with blood flowing through them. Because they are not covered with feathers, in winter they can become damaged. However, a coating of petroleum jelly affords some winter protection.

☙ **During hot weather**, blood flowing through the comb helps to keep a chicken cool.

☙ **Most combs are red**. However, Sebrights have purplish-red combs, while those on Silkies and Sumatras are purple.

Single comb

VARIATIONS IN COMBS

☙ **Single** The most common type, forming a solid but fleshy comb that lies on top of the head. It has five to six points with deep serrations. On males the comb is

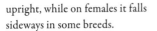

upright, while on females it falls sideways in some breeds.

❧ **Pea** A neat, medium-sized comb with three ridges extending lengthwise from the top of the beak to the top of the head. The middle ridge is slightly serrated and higher than the other two.

Pea comb

❧ **Buttercup** Distinctive and cup-shaped; it starts with a single ridge at the top of the beak, then forms a circle with regularly spaced, antler-like points around it that create a crown.

❧ **Carnation** Sometimes known as a 'King's Comb', this is a single comb with several lobes at its rear.

❧ **Cushion** A small, round and knob-like comb, smooth and positioned slightly above the beak and extending a little way up the head. This type of comb does not have any spikes.

❧ **Horn** Also known as the 'V-shaped' comb, it has two raised and spiky pieces that resemble horns. They are joined at their bases.

❧ **Rose** Tube-shaped, with small, rounded bumps from the top of the beak to the back of the head, where it ends in a backward-pointing spike.

Rose comb

❧ **Strawberry** A low, cushion type, resembling a flat strawberry at the front of the head. The surface is covered in small bumps.

❧ **Walnut** Almost round, this is slightly bumpy with a narrow, crosswise indentation towards its front.

CHAPTER THREE

Keeping Chickens

THEIR RANGE & SIZES

The sizes, colours, shapes and temperaments of the many breeds of chicken vary considerably, with many of these traits being the result of poultry breeders throughout the world developing new breeds.

LARGE OR SMALL?

Chickens can be grouped according to their size, either 'Large-fowl breeds' or 'Bantams'.

❧ **Large-fowl types** can be 'light' or 'heavy'. The light ones, as one might suspect, are lighter than the heavy ones. They may also lay more eggs than the heavy types, which were probably bred mainly for their meat rather than egg-laying abilities. However, light breeds tend to be more nervous, skittish and flighty than heavy types and mainly originated in Mediterranean and northern European regions.

❧ **Bantam breeds** can be either small versions of large-fowl types, or miniatures with no large-fowl counterparts. They are small, inquisitive, colourful and amusing, ideal for keeping where space is limited; they are perfect as pets.

Bantams are less expensive to keep than large-fowl breeds, but their egg-producing abilities are limited.

The classification of a bantam as a true type, or just a size variation of a large-fowl type, often differs from one country to another. Also, some breeds have a wide range of colour variations.

POPULAR BREEDS

The range of chicken breeds is wide and the choice of a particular breed is personal. Popularity also varies from region to region and between countries.

Some breeds are selected for their colour, size or, perhaps, their perceived attitude to life, while others remain popular for their abilities to produce eggs or meat. However, here are 11 general-purpose breeds that will provide both eggs and meat for the table.

❧ **Barnevelder**

❧ **Barred Rock**

❧ **Black Australorp**

❧ **Brahma**

❧ **Buff Orpington**

❧ **Dorking**

❧ **Light Sussex**

❧ **Maran**

❧ **Plymouth Rock**

❧ **Rhode Island Red**

❧ **Wyandotte**

WAYS TO KEEP CHICKENS

Home poultry keepers have a choice of ways to keep chickens, and their selection mainly depends on the amount of land you have available. Even in a small back yard it is possible to have a chicken shed and scratching pen. And remember, chickens make great family pets.

THE 'KEEPING' OPTIONS

❧ **Free-range system**
Traditionally, chickens are kept in this way, able to wander freely during the day over grassed land, but safely secured in a shed, ark or poultry house on wheels at night.

❧ **Fold method** Chickens are kept in a poultry house with an attached wire-netting enclosure in which they can exercise and scratch. There are several variations on the fold system, often using recycled materials bought from reclamation yards.

❧ **Semi-intensive system** Involves allowing chickens to wander freely in a field during the day and be taken into a barn or large shed at night and kept on layers of litter or straw. Chickens have freedom to wander, but this method demands plenty of land and a large shed that can be dedicated to their use.

✤ Deep-litter system Hens are entirely homed in a barn or large barn and never let out. It is a variation on the semi-intensive system, with the chickens kept on regularly topped-up layers of litter or straw, which is completely changed at least annually.

✤ Battery system A solely commercial way of keeping chickens; many people feel that it does not provide chickens with a 'good and natural' existence.

BEING A GOOD LANDLORD

Space is important for chickens, in both their sleeping and day areas.

✤ Sleeping quarters Allow at least 0.028 cu m (1 cu ft) of space for each bird.

✤ Daily scratching and exercising Provide each average-sized bird with 0.03 sq m (3 sq ft). Large chickens need about one-third more space, and bantams one-third less space than an average-sized chicken.

✤ Six hens will provide enough eggs for a family of four to six people.

✤ Three hens are needed to provide eggs for two people.

CHOOSING THE RIGHT SITE

Chickens are susceptible to diseases which are encouraged by damp and airless conditions. In some gardens the choice of position is limited, but often opportunities exist in the orientation of a poultry shed to create privacy and shelter from cold winter winds.

LOCATION, LOCATION, LOCATION!

Finding a position in your garden that ensures your chickens remain healthy is essential. Here are pointers to this success:

❧ Free circulation of air around the building keeps it dry and ensures that the chickens' feet do not remain wet.

❧ Good soil drainage around and under the hen house inhibits the onset of diseases and the possibility of pest infestation.

❧ Avoid positions at the bases of slopes, where trapped air becomes stagnant and cold.

❧ Select a position where sun can keep the building warm and dry.

❧ Ensure positions on the sides of slopes do not enable cascades of water to run directly into the hen house. A drainage trench dug along the top side and filled with rubble helps to prevent this happening.

❧ The closed rear side of a hen house should face away from predominantly cold winds.

❧ Select a position that offers your chickens slight shade during summer.

❧ Always choose a position where summer smells from chickens do not intrude on neighbouring properties.

LOOKING AFTER CHICKENS

If chickens are easy to look after it enhances the pleasure of having them in your garden. Remember to plan for easy access during winter weather.

A convenient supply of clean, piped water is essential, especially if the hen house is not close to a household supply.

The water pipe must be fixed to a stand-pipe, with a tap that does not drip!

Keep all food and equipment in a dry, easily accessible, vermin-proof shed.

CHICK-A-LOO TIME

Apart from the daily collection of eggs, it is necessary to make arrangements for the collection and use of chicken manure.

Each chicken will produce more than 112 g (4 oz) of manure every day. This means that in a single year five chickens produce over 200 kg (440 lb) of manure.

Poultry manure cannot be used immediately. It needs to be added to a compost heap in 2.5–5 cm (1–2") layers, where it will decompose and, after a year, can be dug into soil or used as a mulch.

When cleaning a hen house you may want to take precautions: chicken manure is caustic and may burn your skin. Rubber gloves, overalls and even a face mask and goggles may be needed, especially in hot, dusty weather.

HOUSING YOUR CHICKENS

Getting your chickens accommodated can be straight-forward and easy, especially if you buy 'flat-pack' buildings or, if small, readily assembled ones straight from suppliers. Alternatively, existing sheds can often be inexpensively modified to suit your egg-laying team.

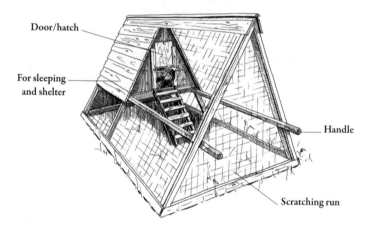

Door/hatch

For sleeping and shelter

Handle

Scratching run

Chicken ark

POULTRY HOUSES TO CONSIDER

❉ **Ark** Ideal where only a few chickens are kept. It is formed of two parts, providing night-time safety and a scratching area during the day. Alternatively, the chickens can be 'free-ranged' during the daytime and put into an ark at night, where they will be safe from predators.

Arks are available in a wide range of sizes to suit 2 to 15 hens. If small, arks can easily be moved by two people.

❉ **Walk-in shed** A shed that has a door, ventilators and 'pop hole' (entrance and exit for chickens) at its front. A nesting box (in which eggs can be laid) is fitted to another side.

Hens are let out during the day, preferably into a wire-netting enclosure that will keep them safe and securely protected.

✤ **Hen house on wheels** This is ideal where chickens are kept as 'free-rangers' in a field. It can be readily moved to fresh ground and, preferably, has a wire-netting enclosure secured to it, if free-ranging is not safe from predators.

When a mobile hen house is old and cannot be readily moved, it can be turned into a static one.

✤ **Static poultry shed and scratching run** Similar to a walk-in shed, but with a large wire-netting scratching run permanently attached to it. It provides ideal protection for your hens in areas where foxes or other vermin are prevalent.

Small chicken house

Compact poultry house

Chicken shed and scratching run

FIXTURES & FITTINGS

To encourage chickens to lay eggs they need to feel secure and have sufficient space during the day and at night (see page 51). Rain-proof conditions are vital, as is good air circulation to prevent the onset of diseases. And they need readily available clean water and nourishing food.

STRUCTURAL FIXTURES

❦ **Roof** If the roof is bare timber, cover it with roofing-felt; ensure that where one sheet overlaps another it has been coated with a sealant to prevent water seeping between them. Additionally, use wide-headed, galvanized felt nails to secure the felt edges to the roof.

❦ **Perches** These are essential to enable birds to sleep comfortably at night. Never overcrowd them and allow at least 20 cm (8") of space for each bird. Use 5 cm (2") thick timber, rounded at its edges, level and positioned about 60 cm (2 ft) above the shed's base. If perches are too high and birds overcrowded, they can damage their feet when jumping down.

❦ **Ventilation** Birds soon suffer from respiratory disorders if their living quarters are not ventilated, but avoid exposure to draughts. Rodent-proof, louvred-types are ideal; alternatively, use wire-netting secured to a window frame, with a shutter that can be opened during the day. Those hinged at their tops are excellent as they prevent rain blowing in.

Nesting boxes These are usually fitted on the outside of one wall: hens can then access the boxes from inside the hen house, and eggs can be collected without disturbing them. Alternatively, boxes can be fitted on the inside. One nesting box is needed for every three hens.

Outdoor poultry feeder

Feeders Dry mash, corn and grit can be given through troughs or self-feeding devices. You can use a small galvanized trough for wet mash.

Single nesting box

Floors These must be dry and, if made of timber, must be vermin-proof. If siting a wooden hen house without a floor on concrete, check that rodents cannot get in between its edges and the wooden framework.

◈

ESSENTIAL FITTINGS

Clean feeders and drinkers are essential; thoroughly wash them when you change the water to prevent diseases spreading to your chickens. When open-topped types are used, there is a risk of food or water becoming contaminated with faeces from wild birds.

Chicken drinker

Drinkers Self-dispensing types are best, as the risk of water becoming contaminated before use is reduced. Low, open types are often used for chicks, but ensure they cannot fall in and drown.

FEEDING CHICKENS

Chickens enjoy eating cooked or raw vegetables (chopped), and leftover cooked pasta and rice. However, young chicks need special food (see page 69), and a balanced diet formed of fats and oils, proteins, carbohydrates, vitamins, calcium, manganese and iodine keeps adult chickens healthy.

A CHICKEN'S DIGESTIVE SYSTEM

A chicken's digestive system is totally different from that in mammals. When a chicken eats food it is rapidly transferred to its gullet (oesophagus), from where it passes into the crop. There, it is usually softened before passing into the gizzard (the equivalent of a stomach in mammals).

This simple, yet successful way of digesting food puzzles many people, with questions such as:

❧ **Can chickens chew?** No. They do not have teeth, instead relying on a strong and hard beak to break up tough and strong food.

❧ **Do chickens have a sense of taste?** Yes. They have glandular taste buds. These are not highly sophisticated, but sufficiently sensitive to enable a chicken to detect what it prefers to eat.

❧ **Why do chickens need grit in their diets?** It acts like teeth when combined with food and makes a chicken's gastric juices more effective.

58

MASH, PELLETS & CRUMBS

❧ **Mash** A balanced food mixture that can be dispensed 'dry' or 'wet'. It is usually given dry when poultry are intensively reared in commercial conditions. In this state, it keeps birds active when there is little else to interest them, but fresh water must be available to them at all times.

Wet mash – a mixture of dry mash and water – is more digestible than dry mash alone. You can also add clean, chopped vegetable waste to it.

❧ **Pellets** Mixtures of food that have been ground and compressed into cylindrical form. They are available from poultry-food stockists and can be given through self-feeding devices if you are away from home most of the day.

❧ **Crumbs** Food that is given in granular form and an excellent way to feed chicks.

❧

AMOUNT OF FOOD

Even within a range of adult birds, the amount of food needed each day varies:

❧ **Large breed** 100–150 g (4–6 oz).

❧ **Bantam breed** 50–75 g (2–3 oz).

HANDLING &
INSPECTING CHICKENS

Chickens are gregarious and need company. Some delight in being handled and talked to in a soft and gentle manner. And the more often they are handled, the friendlier they become. They may also need to be handled to check if they are healthy or in a state to lay eggs.

GETTING TO KNOW CHICKENS

At first, chickens may be wary of you, but when offered a handful of food pellets, they quickly assume you are friendly. Usually, the chicken highest in the pecking order will be the first one to come to you. If, however, you initially pick up one that is lower down in the hierarchy, those higher up may become jealous and walk over to check what you are doing and to make their presence known.

PICKING UP A CHICKEN

Unless a chicken stands immediately in front of you and allows itself to be picked up, it needs to be carefully ushered into a corner. It is then possible to bend down and slowly pick it up, hugging its wings close to its body; hold the bird securely.

Take care not to panic it, as this may temporarily reduce its egg-laying ability. Additionally, if the bird is panicked it may suddenly flap its wings and try to get away from you, falling to the ground and possibly injuring itself.

CARRYING A CHICKEN

Put one hand under its rear end and tuck the head under your arm, with it slightly lower than the rear part. Some poultry enthusiasts like to have the head pointing forward, but this enables a chicken to see what is happening and may frighten it.

INSPECTING A CHICKEN

Regularly checking a chicken is essential to ensure good health.

Checking the bird's crop Best undertaken at dusk, but when the light is still good. If the crop is full, you are providing the right amount of food. If not, give added rations.

Checking for broodiness A broody hen fluffs up her feathers, squawks and is reluctant to leave a clutch of eggs. Broody hens can be isolated in a coop with a slatted front and within sight of other hens. After a few days they can be put back with the other hens.

Looking for external parasites Draw apart the feathers and check their undersides. Also, look for bare areas of skin, as this is a clear indication they are infested; also check legs for scaly leg mite, when the legs thicken, with raised, scaly, crusty ridges; treat with an appropriate dust or spray.

NEVER...

**Pick up a chicken by its feet or neck, as this causes severe damage, from which it may not fully recover.

**Pretend to hypnotize a chicken by placing it on its back – the bird may suffer heart failure, especially if a heavy breed.

KEEPING A COCK-BIRD

Keeping a cock-bird (also known as a cock or rooster) in with your hens is not essential for the production of eggs. Also, there is no evidence that the presence of a cock-bird encourages hens to lay more eggs. However, should you wish to breed from your hens, of course a cock-bird is vital.

TO HAVE OR HAVE NOT?

For breeding purposes, it is essential to have a cock-bird bought from a reputable source to ensure his pedigree is known. If kept for fertilizing hens, he should not be closely related to them.

There are advantages and disadvantages in keeping a cock-bird, and these include:

Advantages

✾ He will keep your hens under control, preventing them squabbling and ensuring they stay placid.

✾ The longer he is kept among your hens, the greater the number of fertile eggs you will have; this is important if you wish to raise chicks.

White Crested Black Polish

Disadvantages

❧ In suburban gardens, a cock-bird's crowing each morning may irritate neighbours and cause you problems.

❧ A cock-bird introduced into a group of hens that have never seen one before can disturb them and put them off laying eggs.

Bantams

❧ 'Rescued' hens (see page 27) are especially agitated at the sight of a cock-bird and will certainly not be used to his attitude to life and his sexual demands.

❧ If you only want to have eggs for eating, the cost of buying and feeding a cock-bird is unwarranted.

❧ Eggs may be fertilized and some people do not like eating these, even though they are not harmful.

❧ Cock-birds may need to be de-spurred to prevent them damaging hens when mating.

COCK-FIGHTING

In mediaeval times, the morality of inflicting suffering on animals was not questioned, with both women and men enthusiastically watching bear-baiting and bull-baiting. Boys and girls traditionally organized cock-fighting on Shrove Tuesday, with adults making bets on family-owned birds.

Children also arranged cock-throwing, where sticks and stones were thrown at tethered chickens, killing them and taking them home for supper. However, to prolong the 'sport', the bird's feathers were often greased to make sticks and stones glance off.

KILLING CHICKENS

The pleasure in keeping chickens is only clouded when one
of them has to be killed, perhaps through ill-health or when aged.
This is part of nature and has to be accepted, but terminating the
life of a bird must be carried out according to regulations and
without causing undue stress and suffering to it.

IS IT LEGAL?

❧ **You can kill** your own birds (on
your own land) or arrange for your
chickens to be killed in an approved
slaughter house. It is essential to kill
chickens in the correct way.

❧ **You can eat meat** from your own
chickens, but they must have been
slaughtered humanely and legally.

❧ **You can't sell meat** from your
own chickens that you have killed.
Only if the animal was killed in an
approved slaughter house would
this be possible.

Also, only if a chicken were killed in
an approved slaughter house would
it be legal to feed the meat to paying
guests, e.g. in a B&B enterprise.

TERMINATION

There are several commercial methods of killing a chicken, but for the home poultry keeper dislocation of the neck is the best. It causes concussion and rupturing of the spine and, when undertaken correctly, the bird immediately loses consciousness. The method is:

1 Select the chicken and place it in a large box in a quiet, slightly dark shed away from your other birds. Cover the box with a sack until you are ready to despatch the chicken.

2 Hold both feet of the bird securely with your left hand and raise it to about shoulder height.

3 Use your other hand to clasp the bird's head and turn it at a right-angle to the neck.

4 With a sudden pull downwards, snap the neck. If the neck parts from the body, this is not a problem, but can be rather messy!

5 The bird usually flutters its wings for a few moments.

6 Suspend the bird from a strong hook in a cool, shaded, vermin-proof place.

TRAINING

Before attempting, for the first time, to end the life of a chicken, it is wise to attend one of the many training courses aimed at the novice chicken-keeper, to ensure you cause no undue stress at its death.

DISPOSING OF A DEAD CHICKEN

The body of a dead chicken must be disposed of in a legal manner. It cannot just be put in a bag and left for a local waste-disposal company to take away.

Place the dead chicken inside two plastic bags and, temporarily, put it in a shed. Then contact your local Animal Health Office for advice about disposal of the body. This also applies to chickens killed in road accidents, as well as those that die through old age or infirmity.

PREPARING A CHICKEN FOR COOKING

Preparing a chicken for cooking involves plucking, hanging, drawing and trussing. However, there are many different opinions about how best to carry out these procedures; here are some guidelines.

TERMINOLOGY & TECHNIQUES

Plucking Removal of the bird's feathers in preparation for cooking.

❧ A chicken's body remains warm and floppy for about 45 minutes after death. During this period, feathers normally come out easily and most keepers believe this to be the best time for plucking.

❧ Plucking can also be undertaken when the body is totally cold.

❧ On a windless day, spread a sheet on the ground and suspend the chicken by its legs from a strong

66

hook. Alternatively, sit on a chair with the bird on your lap.

🐾 Start by plucking the long, stiff flight feathers. Hold several together and give a sharp tug.

🐾 Next, pluck feathers from the legs, then the body.

🐾 Eventually the body will be free from feathers. Then cut off the bird's neck at the position of dislocation.

Hanging The flavour of a chicken's flesh is improved by 'hanging'; it enables fats in the tissue to become firm.

🐾 Either suspend the chicken in a clean, vermin-proof, well-ventilated shed with a temperature under 3°C (37°F) for 24 hours or place it in a refrigerator.

Drawing The removal of the bird's innards is known as evisceration. A clean table and knives are essential for this task.

🐾 Use a sharp knife or poultry secateurs to cut the neck about 2.5 cm (1") above the shoulders.

🐾 Use your fingers and a knife to pull away the crop (where, after being eaten, food is stored until it passes into the stomach) and cut close to the neck cavity.

🐾 Turn the body around and use a sharp knife to sever the fatty oil sac, sometimes known as the parson's nose. If left, it may give the flesh a peculiar taste.

🐾 Carefully cut between the earlier position of the parson's nose and the vent (the orifice from which the chicken defecates, urinates and expels eggs). Take care not to pierce the rectum, then pull this out, complete with the intestines.

🐾 Remove the rest of the innards, which includes the heart, kidneys and liver.

🐾 Lightly wash the inside of the chicken, then dry thoroughly.

Trussing The purpose is to secure loose flesh over the body and hold the legs in place, so that the bird can be easily cooked.

RAISING NEW CHICKENS

There are several ways to acquire hens and to start keeping chickens. The range of large fowls and bantam breeds is wide and the number of eggs they lay within a year varies hugely. Local chicken suppliers will be able to suggest the best breed for your area and climate.

WAYS TO START

✦ Buy hens that are either currently laying eggs or are about to start (these are known as 'Point-of-Lay' (POL) pullets).

✦ Buy young chicks that you can raise until they become pullets and start to lay eggs.

✦ Keep a cock-bird with your egg-laying team and produce fertilized eggs, which can be encouraged to hatch. Usually the mother hen will attend to this, using her own body temperature of about 37.7°C (100° F) for about 21 days.

IDYLLIC & MEMORABLE

Few occasions when keeping chickens compare with the moment a young chick emerges from its egg. Smiles, especially on the faces of young children, gladden the moment and make it even more memorable. If a mother hen is present, she can be left to raise them, but the chicks will need special food until about 16 weeks old.

❧ **For the first 24 hours** no food is required during this period.

❧ **Then, until 6–8 weeks old,** give high-protein proprietary food, usually known as 'starter crumbs'. Initially, use milk to slightly dampen them; cod-liver oil can be added in small amounts.

❧ **At 6–8 weeks** slowly change their food to 'grower crumbs', but at first gradually combine them with 'starter crumbs', their earlier food.

❧ **At 16 weeks** gradually introduce the chicks to adult food.

CHICK FEATHERING

At 6–8 weeks old, chicks start to lose their initial fuzzy coverings and develop early feathers. This change of feathers is known as 'feathering up' or 'feathering out' and, when complete, the chicks have 'hardened off'.

KEEPING CHICKENS AS PETS

Ten or more years ago few people would have considered keeping chickens as pets, although some breeds have long been known for their friendly natures and quick responses to personal names. Dogs and cats are traditionally pets, but chickens also have a great deal to offer families.

Light Brahmas

POPULAR CHOICES

Up to three million chickens are now kept in back gardens in the British Isles, primarily for their eggs, but also as pets.

❧ Some chickens are bossy, others demure and restrained, but as pets they have the bonus of providing you with food – primarily eggs, as few people would like to eat their family pet!

❧ Chickens are gregarious and, to keep them happy, a group of four or five is best.

❧ Never keep a chicken on its own, as companionship from another bird is vital.

❧ Most chickens live four to seven years, with their egg-laying abilities decreasing after the first year (depending on the breed).

Whereas hens that are normally kept for laying eggs are killed when their egg-laying abilities radically diminish, as family pets they are usually retained until infirm and it becomes a kindness to despatch them humanely.

BREEDS AS PETS

Both large-sized breeds and bantams are ideal as pets.

Large-fowl breeds

- **Barnevelder**
- **Brahma**
- **Dorking**
- **Faverolles**
- **Orpington**
- **Plymouth Rock**
- **Rhode Island Red**
- **Silkie**
- **Sussex**
- **Welsummer**

Bantams
These can be smaller versions of large-fowl breeds or 'true' bantams. True bantams include:

- **Rosecomb**
- **Pekin**

Black Rosecomb Bantams

STRESS THERAPY — FOR BOTH OF YOU

Chickens delight in being picked up, held securely in your arms and stroked. It is good for them and stress-reducing therapy for you.

- Begin stroking chickens early in their lives – and talk to them. If you are fortunate enough to have chicks, stroke them while still young – but don't let small children do this, as they may inadvertently harm them.

- Chickens can be jealous and even vengeful if affection is solely given to one hen in their group. Therefore, treat all of your chickens the same.

KEEPING CHICKENS HEALTHY

Whenever animals are kept relatively close together there are increased risks of pests and diseases affecting them. 'Problem habits' and 'stress-related disorders' may also occur. However, most chickens, if given wholesome food, clean water and sufficient space, will remain healthy.

SIGNS OF GOOD HEALTH

Good health in chickens can be seen in several ways:

🐓 They are calm, but not silent.

🐓 Moving around normally.

🐓 Feeding and drinking as usual.

🐓 Preening regularly.

🐓 Sunbathing or dust-bathing.

🐓 Perching with confidence.

🐓 Putting on weight (if young and up to the age of 18 weeks).

🐓 Laying eggs (if a hen and of the correct age).

🐓 Sparring or mock fighting (in young birds).

🐓 The bird's droppings should be dark, firm and have a white tip.

꿍

PREVENTION OF PESTS & DISEASES

It is easier – and less expensive – to prevent disease problems than to have to deal with them when they reach epidemic proportions. Here are a few precautionary measures:

🐓 Keep living areas clean.

🐓 Give a correct diet for the chickens' age.

🐓 Daily check water dispensers or bowls to ensure they are not contaminated with faeces or organic material.

🐓 Feed with fresh kitchen leftovers (avoid those from other households).

✤ Do not cram a large number of chickens into a small area, during the day or at night.

✤ Store food in clean containers in a dry, vermin-proof shed.

✤ If you do notice a chicken that is ill, immediately isolate it and seek veterinary advice.

❧

PESTS TO CONSIDER

There are two types – internal and external. There are various poultry treatments available to deal with them. Consult your vet as to which is the most safe and effective.

Internal parasites

Try to avoid worm infestation by regularly checking your flock, keeping all equipment clean and moving your hens on to clean ground regularly. You can also worm twice a year or use a herbal preventative like apple cider vinegar in their water – 1 teaspoon per litre, provided the drinker is plastic.

✤ **Caecal (cecal) worms** These are yellow-white to white worms up to 18 mm (¾") long. Severe infestations cause inflammation of the caeca (blind gut), resulting in pale birds with a drooping nature. Birds huddle together, appetites diminish, they drink less water and become emaciated, often accompanied by diarrhoea.

✦ **Gape worms** Occasionally known as forked worms, they are red, Y-shaped and 6–18 mm (¼–¾") long. Birds gape, opening their mouths without making any noise. Additionally, they shake their heads, have difficulty in breathing and lose their appetite.

✦ **Roundworms** Oval, thick, yellowish-white and up to 10 cm (4") long; they can be seen in the droppings of infected chickens. Severe infestations cause listlessness, poor growth and diarrhoea. Young birds, if not treated, may die.

✦ **Tapeworms** Flat and segmented bowel parasites; each part is a developing egg-case and, when mature, may break off and pass out of the bird. This contaminates land and perpetuates the problem. Chickens lose weight, become lethargic and have difficulty in breathing. Free-ranging chickens are more likely to become infected than those solely kept in runs.

✦ **Threadworms** Minute, thread-like and sometimes known as capillary worms. There are several types, each producing slightly different symptoms, but usually causing weight-loss and diarrhoea. Severe infestation can cause death.

External parasites
✦ **Fleas** Small, able to jump, and with skin-piercing mouth parts that suck blood and transmit diseases. They are found in clusters on the skin, hidden away and cause restlessness and itching; birds often peck at themselves.
Treatment: Remove chickens from their enclosure, clean it thoroughly and replace with fresh bedding. Dust infected birds with a flea powder and check every few weeks for any further infestation.

✦ **Ticks** Small, tough and resilient, they invade poultry at night, resulting in areas of bare skin, irritation and anaemia. Egg production decreases.
Treatment: Remove all poultry from the enclosure, wash the inside of the house thoroughly, then use a recommended insecticide on both house and poultry.

✦ **Lice** Usually 3 mm (⅛") long and ranging from off-white to brown and dark grey. They feed on dried skin and feathers, irritating the chicken which becomes drowsy. Wings sag and egg production diminishes.
Treatment: Remove the chickens from their enclosure, thoroughly clean the chicken house and treat with a recommended insecticide.

Mites More pernicious than fleas, they feed on the blood of chickens, mainly at night. They cause severe skin irritation, often resulting in anaemia and sometimes death.
Treatment: Remove chickens from their enclosure, take out and burn the bedding and thoroughly clean the building. Then dust chickens and house with an insecticide.

~

DISEASES OF CHICKENS

The novice chicken keeper needs to be aware of two highly infectious diseases: Newcastle Disease (Fowl Pest, with symptoms including sneezing, listlessness, gasping for air and green diarrhoea) and Avian Influenza (Bird Flu). These are highly contagious – contact your vet and your local Animal Health Office as soon as they are identified.

~

STRESS-RELATED PROBLEMS

Chickens thrive in a quiet, restful and unchanging routine, where food arrives at the usual times and, preferably, given by the same person.

Symptoms of stress are varied and include:

- Birds become 'flighty' and act in an unpredictable manner.

- Diarrhoea.

- Flinching and shying away.

- Laboured and irregular breathing.

- Lethargy.

- Reluctance to eat and drink.

- Shaking head vigorously.

- Those at the bottom of the pecking order (see page 30) may be forced away from food and water.

Stress-preventative measures include:

- Regular feeding.

- Never putting an excessive number of chickens in a hen house or enclosure (see page 51).

- Sufficient feeders and drinkers to enable all chickens to eat or drink at the same time.

- Avoiding loud noises.

- Keeping enclosures cool during summer and warm in winter.

OTHER PROBLEMS

These are not normally initiated by pests and diseases, but are difficulties chickens sometimes encounter.

❧ **Bumble foot** Usually results from a foot becoming cut, the wound healing on the outside, but still being infected internally and preventing the bird walking properly. Seek veterinary advice to cleanse the infection.

❧ **Cannibalism** Primarily caused by overcrowding (see page 51 for optimum space). Birds peck at each other's feathers and flesh; the problem is easier to prevent than to control. Insufficient food or a radical change in diet can also trigger this problem.

❧ **Crop impaction** Sometimes known as 'crop binding', this occurs when a bird's crop is packed with food that doesn't easily pass into the gut. Giving the bird a drink of warm water distends the crop, which can then be massaged to ease the blockage. Alternatively, hold the bird upside down and gently squeeze the crop. Afterwards, give water, but not food, for 24 hours.

❧ **Egg-eating** Usually begins by accident, when an egg breaks and a chicken pecks at it. It tastes good and other hens join in. Can also occur when hens are congested. Treatment includes giving them more space, further calcium in their diet and suspending bunches of green leaves to peck at (it takes their attention away from eggs).

76

Egg-binding An egg-laying disorder when an oviduct (the long tube where eggs form, ready for laying) is too small. Can also be caused by a broken egg stuck in the oviduct. Give the hen olive oil and place her in a straw-lined box in a warm, quiet area for 24 hours.

Feather-pecking Areas around the rump, back and tail are pecked by other birds, sometimes leading to cannibalism. It is usually associated with hens kept in runs, rather than as free-rangers. Remove seriously infected birds and place them in a clean, straw-lined box. If the problem continues, their beaks may require clipping.

Internal laying A serious problem, occurring when an egg takes the wrong channel inside a hen and becomes infected. Little can be done to help her and she is best killed.

Prolapse Sometimes known as 'down behind', this happens to hens during and after their second year of laying. A portion of the oviduct (egg-laying channel) protrudes from the hen's rear end. This can encourage cannibalism. Overweight hens and excessive straining to lay an egg create the problem. Wash the area, coat it in olive oil and use a finger to push the oviduct back inside. Then, place her in a clean, straw-lined box in a warm, quiet area for a week. Reduce the food given to her, but make water readily available.

CHAPTER FOUR

Chickens in Superstitions

SUPERSTITIONS

Chickens have been kept for their eggs and meat for thousands of years and during this time many superstitions have arisen about them. These are usually steeped in country folklore, some amusing, others questionable, but all richly part of our heritage. Here are a few of them.

In Europe it was earlier thought to be unlucky if a hen laid an even number of eggs, and should this occur, one should be removed or no chicks would hatch from them.

If all the eggs laid at a single laying happened to produce cockerels, this was thought to be lucky.

If a hen (not a cock-bird) crowed, this was said to be a prelude to the coming of evil.

78

It was said that immediately after the death of a farmer all his chickens would roost by midday, rather than in the evening.

The setting of eggs (putting them in a position where they can hatch) had fertility taboos for country people until the late 19th century. It was thought unlucky to carry fertile hens' eggs over running water, or to position them to 'set' on a Sunday or in the month of May (considered to be the month of witchcraft).

A Persian superstition suggests that cock-birds should crow at nine and

twelve in the morning and at night. If they do this their owners can expect sudden good fortune.

Housewives would pass a lighted candle over set eggs, or make crosses on them, to save them from foxes and weasels.

Spring flowers were said to influence the number of eggs that would hatch. If a posy containing fewer than 13 primroses was taken into a house, the number of eggs that hatched would be indicated by the number of flowers in the posy.

It was considered unlucky to sell eggs or to take them indoors after sunset.

The persistent crowing of a cock-bird between dusk and midnight is said to foretell a death.

A cock-bird crowing outside a back door was said to foretell a stranger visiting the housewife.

In Indonesia, during a Hindu cremation ceremony a chicken was used to channel away evil spirits. It was tethered by a leg and kept present throughout the ceremony, enabling malevolent spirits to enter the chicken rather than family members. After the ceremony, the chicken was returned, unharmed, to its normal way of life.

In earlier years, to attract a husband a maiden would hang a dried wishbone from a chicken or turkey over the entrance to her home.

❧

In the Voodoo religion, a chicken's foot held great power. Voodoo practitioners used them as charms and talismans to ward off evil spirits.

❧

In ancient Greece, birds were sacrificed, but chickens were normally exempted from this as they were considered to be exotic and highly respected animals.

In Central European folklore the devil is claimed to flee at the first crowing of a cock-bird.

❧

An early Roman superstition was that if a chicken refused to eat food offered to it, it was a bad omen for the immediate future.

❧

The American baseball player Wade Boggs, who played with the Boston Red Sox and New York Yankees, would eat a meal of chicken before every game. He claimed this routine helped him amass 3,000 career hits.

The 'wishbone' of a chicken or turkey can be broken to encourage a wish fulfilment. Usually this is done at Christmas or Thanksgiving. The breaking ritual began with setting aside the wishbone for three days to allow it to dry and become brittle. At this stage – with a happy result anticipated – it is called a 'merry thought'.

The breaking ritual is performed by two people, each person wrapping their little finger around one side of the wishbone. Before proceeding, each of the participants makes an undisclosed wish. When an agreed signal is given, each person tugs and uplifts their side of the wishbone; it usually snaps. The person with the centre piece attached to the part they hold gets their wish fulfilled.

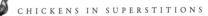

After eating a hard-boiled egg, it is essential to tap a hole though the bottom of the shell. If this is neglected, witches will go to sea in it and cast bad luck on sailors and sink ships.

ROOSTER STAR SIGN

The chicken, in the form of a rooster, is a Zodiac sign in the Chinese calendar. Someone born under this sign often has a quixotic nature, being a dauntless hero, but often misunderstood. Can you see the Rooster in your life from the following traits?

❧ Outwardly, the epitome of self-assurance, but at heart conservative and old-fashioned.

❧ Appearance is very important and, although usually knowledgeable, he or she would rather be judged by appearance than by his or her intelligence and wisdom.

❧ Always seeking improvement, never satisfied with his or her current performance.

❧ In a crisis, a person who is resourceful, talented and self-reliant.

❧ A questioning nature, requiring proof of a fact or activity.

❧ Independent and relying on himself or herself for moral support and solutions to problems.

CHAPTER FIVE

Slang, Proverbs, Sayings & Phrases

CHICKEN SLANG

Chickens are so popular and widely kept that they have engendered a rich pageant of slang terms. Some of these are flattering, others not, but they are all a reflection of life and of how, when in general use, language is rich in metaphor. Here is a range of chicken slang.

* **Bad egg** – a person who is less than honest and has poor moral standards.

* **Brood over it** – to worry and consider a problem thoroughly.

* **Chick** – a young girl.

* **Chick-feed** – a trivial amount of money.

* **Chickadee** – North American song bird; also used when referring to pretty girls.

Chickaleary – artful.

Chicken – afraid.

Chicken feed – a paltry sum.

Chicken-fixing – a nautical term used by sailors in the 1880s when a name could not be remembered.

Chicken food – an early naval term sometimes given to unappetising food rations.

Chicken-hammed – being bandy-legged.

Chicken-hearted – timid and cowardly, with a weak will.

Chicken-in-the-clay – a dead fowl rolled in mud and then roasted in preparation for eating.

Chicken-livered – faint-hearted and cowardly.

Chicken nabob (1811 term) – a person who returns from the East Indies with a fortune of £50,000 or £60,000.

Chicken perch – rhyming slang for church.

Chickened-out – gave up.

Chickens have come home to roost – sins have found out a person, who is then in disgrace.

Chickery-pockery – dishonest and shameful dealings.

Cock-and-bull story – elaborate and fanciful lie.

Cock crows best on his own dung heap – confident of his surroundings.

Cock-eye – turned or twisted towards one side.

Cock-eyed – squinting or cross-eyed.

Cock Lane Ghost – a tale of terror without any truth; an imaginary tale of horrors.

Cock of hay (also, haycock) – small heap of hay thrown up temporarily, to be removed later.

Cock of the walk – the boss or a bully.

Cockpits – squalid areas in which cockfights took place. However, during the First World Way (1914–18), and since, pilots have adopted this name for the cramped quarters in which they fly an aircraft.

Cocksure – bragging.

Egg on your face – caught telling an untruth. Alternatively, when something turns out to be wrong in an embarrassing manner.

Empty-nest syndrome – depression and loneliness when children have left home.

Feather your own nest – to abuse one's position of power in order to accumulate personal wealth.

Feeling broody – wishing to be pregnant and having young.

Flown the coop – gone.

Fussing like an old hen – very pernickety.

Go to bed with the chickens – go to bed early.

Hens and chickens – a popular term in the 1850s for small and large pewter pots.

Hen-frigate – nautical term for where the captain's wife rules the ship.

Hen house – a place where the woman rules.

Hen house – brothel.

Hen house – North American term for the women's part of a prison or other area of confinement.

Hen party – a female get-together as a counterpart to the men's stag night before a wedding.

Hen-pecked – nagged.

Hen-roost – a gallery in an old chapel, usually in a college, in which the masters' wives sat.

Hen-toed – walking with ones feet turned inwards.

Like a chicken with its head cut off – running around in no specific direction.

Live like fighting cocks – to live in luxury, well-fed and cared for.

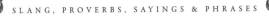
❧ **Mother hen** – a loving and caring mother.

❧ **Nest egg** – some money put aside for the future.

❧ **No spring chicken** – being old.

❧ **Pecking order** – your place in society and among other people.

❧ **Put up a squawk** – argue.

❧ **Raise your hackle feathers** – be visibly annoyed.

❧ **Ruffle your feathers** – find that something annoys you.

❧ **Rule the roost** – be the boss.

❧ **Scarce as hen's teeth** – impossible to find.

❧ **Scrambled egg** – refers to the irregular banding on hats of officers above a certain rank in the military services.

❧ **Scratching a living** – earning just enough money to get by.

❧ **Something to crow about** – having exciting news.

❧ **'Strutting' your stuff** – showing off your talents (like a cock-bird).

❧ **Sunny-side up** – a happy attitude to life.

❧ **That cock won't fight** – dodges the question and won't answer.

❧ **The rooster may crow** – but the hen delivers the goods.

❧ **The rooster may rule the roost, but the hen rules the rooster.**

❧ **To cry cock** – to claim victory or to assert oneself to be superior.

❧ **Up with the chickens** – awaking early with the sunrise.

❧ **Walking on egg shells** – treading softly and carefully to avoid upsetting anyone.

PROVERBS, SAYINGS & PHRASES

These are not always kind, but are usually perceptive and reflective of life and its daily happenings. Many are associated with chickens and a few of them are given here.

❧ A large cock does not suffer a small one to crow.

❧ Black hens lay white eggs.

❧ Chickens and children must always be hungry.

❧ Don't cackle if you have not laid.

❧ Don't count your chickens before they are hatched.

❧ Even clever hens sometimes lay their eggs among nettles.

❧ Every cock crows on its own dunghill.

❧ Every cock scratches towards himself.

❧ Fat hens lay few eggs.

❧ He who feeds the hen ought to have the eggs.

❧ Hens prefer to lay eggs where they see an egg.

❧ If a hen hadn't cackled, we should not know she had laid an egg.

❧ In cold weather cocks crow at midnight.

❧ It is bad when a hen eats at your house and lays at another.

❧ It is not the hen that cackles most that lays the most eggs.

🐓 Prepare a nest for a hen and she will lay eggs for you.

🐓 The cock often crows without a victory.

🐓 The cock shuts his eyes when crowing because he knows it by heart.

🐓 The cock that sings untimely must have its head cut off.

🐓 The hen ought not to cackle in the presence of the cock.

🐓 The hen that stays at home picks up the crumbs.

🐓 You can't make an omelette without breaking eggs.

∾

CHICKEN CURRENCY!

Few people could have imagined that chickens would be used as payment in brothels, but it did happen. Or, to be precise, in a later development of a brothel that opened in La Grange, Texas, in 1844. During the Great American Depression (mainly the 1930s), patrons traded chickens for the brothel's services.

CHAPTER SIX

Breeds & Fame

FAMOUS ASSOCIATIONS

From time immemorial chickens have been prized for their eggs, meat, decorative qualities and 'sporting' abilities. New breeds were developed, many of which have become popular and well-known. A few have also achieved fame through the people who took a particular liking to them.

Alert and active game birds

Black Minorca

🐓 **Thomas Edward Lawrence**
(1888–1935), popularized by the
American writer and broadcaster
Lowell Thomas as Lawrence of
Arabia, was especially fond of the
Minorca breed (earlier known
as the Red-faced Black Spanish).
Lawrence was an idealist, a leader,
strategist and thinker who was
misled by the English and French
governments during the First World
War, much to the later detriment of
the Arab peoples.

🐓 **Gregor Johann Mendel**
(1822–1884), an ethnic German
born in Austria, founded the science
of genetics. Part of his studies
involved Andalusian chickens, an
old Mediterranean breed, in his
hereditary experiments. If a blue
male and a blue female mate, the
resulting offspring are 50% blue,
25% black and 25% splash (silver-
white, with splashes of blue). When
breeding from a black and splash, the
result is usually 100% blues. But this
result is only reliable if a blue-bred
black is used.

Silkies

Major F. T. Croad famously introduced into England in 1872 a breed from the Langshan District (just north of the Yangtse-Kiang River) in China. However, it was Major Croad's niece, Miss A. C. Croad, who was mainly responsible for establishing the breed in Britain. It was introduced into North America in 1878, and in 1879 the breed was taken to Germany, becoming known as the German Langshan.

Marco Polo (c.1254–1324), the famous Venetian merchant and traveller, mentioned 'furry' chickens (now known as Silkies) in China and at one time they were thought to be crosses between rabbits and chickens. The Renaissance Italian naturalist and writer Ulisse Aldrovandi (1522–1605) described them as 'wool-bearing' chickens and 'resembling black cats'. They are now popular in many countries and are often kept as family pets.

The Coronation variety of the Sussex breed was developed for sale at the time of the anticipated coronation of King Edward VIII in 1936. However, this did not happen as he abdicated and his younger brother became King George VI.

Queen Victoria (1819–1901) had a passion for dogs, particularly a Cavalier King Charles Spaniel named Dash, but she also kept chickens. Many breeds of chicken were given to her, including Brahmas, Cochins and Dorkings.

🐾 **William Cook** (1849–1904), originated the now world-famous Orpington breed in 1886 by crossing Minorcas, Langshans and Plymouth Rocks. At that time William Cook was living in the village of St Mary Cray, near Orpington in Kent. In 1895 the breed was shown in Madison Square Gardens, New York; its popularity then soared.

🐾 **Sir John Saunders Sebright** (1767–1846), Member of Parliament for Hertfordshire, England, was an agriculturalist and writer especially known for breeding cattle, pigeons and chickens. He spent more than 30 years creating the Sebright breed, a small, bantam-type chicken with laced plumage. It is similar to the Polish breed and was produced by crossing Hamburg, Nankin and Polish birds. It was the first poultry breed to have its own specialist club.

Silver Sebright Bantams

INDEX

First published in the UK in 2012 by
Green Books
Dartington Space, Dartington Hall,
Totnes, TQ9 6EN, UK

ISBN 978 0 85784 092 9

1 3 5 7 9 10 8 6 4 2

PRODUCED BY
Fine Folio Publishing Limited
6 Bourne Terrace, Bourne Hill, Wherstead,
Ipswich, Suffolk, IP2 8NG, UK

DESIGNER
Glyn Bridgewater

EDITOR
Mandy Greenfield

PICTURE CREDITS
The British Library: pages 80, 82 and 89
Front cover: Fred Geary,
The State Historical Society of Missouri

Printed on Munken Print Cream
paper and bound by
T J International
Trecerus Industrial Estate
Padstow, Cornwall, PL28 8RW